Natural Resource Management in Dryland Agriculture

The Authors

Dr. D.H. Ranade is Graduate in Agricultural Engineering from JNKVV, Jabalpur, post graduate in Soil and Water Conservation Engineering from I.I.T. Kharagpur and doctorate in Civil Engineering from DAVV, Indore. He has more than three decades of experience in research and extension of natural resource management aspects particularly in rainfed agricultural system. He is a recipient of ICAR, New Delhi's prestigious "Dr. Vasantrao Naik award 2009" for his outstanding research in the field of rainwater and dryland agriculture. At present, he is looking after the research activities in All India Coordinated Research Project on Dryland Agriculture, College of Agriculture, RVSKVV campus Indore as Chief Scientist.

Prof. Santosh Mujalde is Graduate in Bachelor of Agriculture from IGKVV, Raipur (Chhattisgarh), post graduate in Agronomy from CCSHAU, Hisar and doctorate it is in progress in Agronomy from Bangor University, UK. He has more than ten years of experience in research and extension of natural resource management aspects particularly in rainfed agricultural system. He is a recipient of Overseas Scholarship for Higher Education from Govt. of Madhya Pradesh during 2010 to 2013 and International Student Ambassador Award from Bangor University, UK 2012. At present, he is looking after the research activities in All India Coordinated Research Project on Dryland Agriculture, College of Agriculture, RVSKVV campus Indore as Scientist (Agronomy) and Officer In-charge Farm of Agriculture Research Farm RVSKVV campus Indore since February 2015.

Dr. Indu Swarup is Graduate in Bio-Science from Rani Durgawati University, Jabalpur, post graduate in Plant Breeding and Genetics from JNKVV, Jabalpur and doctorate in Genetics from Barkatulla University, Bhopal. She has more than three decades of experience in research, extension and teaching of Plant Breeding and Genetics aspects particularly in rainfed agricultural system. At present, she is looking after the research activities in All India Coordinated Research Project on Chickpea, College of Agriculture, RVSKVV campus Indore as Senior Scientist.

Dr. Akhilesh Singh born in 1975. He did B.Tech in Agricultural Engineering in 1999 from Allahabad Agricultural Institute, Allahabad the author completed M.Tech. and Ph.D. in Soil and Water Engineering in 2002 and 2009, respectively from JNKVV, Jabalpur. He worked as SRF and RA for more than 7 years at College of Agricultural Engineering, JNKVV, Jabalpur. Presently he is working as Scientist (Agril. Engg.) at RVSKVV, Gwalior from 201. He has published more than 15 research papers in NAAS rated journal, 12 Technical bulletins/ manuals/ practical manuals, one ISBN edited book.

Dr. D.V. Bhagat is Graduate in B.Sc. Agriculture and Post Graduate in Agronomy from J.N.K.V.V., Jabalpur (MP) and doctorate in Life Science (Botany) from A.P.S.V.V., Rewa (MP). He has more than three decades of experience in research and extension of resource management aspects. He has received *"Bharat Shiksha Ratan"* award on 7th April, 2018, from Global Society for Health and Educational Growth, New Delhi. At present he is looking after the research activities in All Indian Coordinated Research Project on Dryland Agriculture, College of Agriculture, R.V.S.K.V.V. campus, Indore as Senior scientist.

Dr. O.P. Girothia is Graduate in B.Sc. Agriculture and Post Graduate in Agronomy from J.N.K.V.V. Jabalpur (MP) and doctorate in Agronomy from Mahatma Gandhi Chitrakoot Gramoday Vishwavidyalaya, Satna (MP). He has more than three decades of experience in research and extension of resource management aspects particularly in *rainfed* agricultural system. He has received *"Bharat Shiksha Ratan"* award on 7th April, 2018, from Global Society for Health and Educational Growth, New Delhi. At present he is looking after the research activities in all Indian Coordinated Research Project on Dryland Agriculture, College of Agriculture, R.V.S.K.V.V. Campus, Indore (MP) as Senior Technical Officer (T6).

Natural Resource Management in Dryland Agriculture

– *Authors* –

D.H. Ranade

S. Mujalde

Indu Swarup

Akhilesh Singh

D.V. Bhagat

O.P. Girothia

BIOTECH

2019

BIOTECH BOOKS®

ISBN 978-81-7622-442-0

Published by: **BIOTECH BOOKS®**
4762-63/23, Ansari Road, Darya Ganj,
New Delhi - 110 002
Phone: 09899295259
E-mail: biotechbooks@yahoo.co.in

Typeset at: **Classic Computer Services**
Delhi - 110 035

Printed at: **Replika Press Pvt. Ltd.**

PRINTED IN INDIA

Preface

The **College of Agriculture, Indore** campus of Rajmata Vijayaraje Scindia **Krishi Vishwa Vidhyalaya, Gwalior** being one of the centers, have been working on various dryland research programmes in the farmer's fields directly through **Operational Research Project for Dryland Agriculture,** in Vertisols of Malwa region. At this centre, under this project, attempts have been made to conserve the natural resources for sustaining the agricultural productivity in the area. Many significant achievements have been made by the team of the scientists working at this centre particularly for the crop improvement and agronomical technology for dryland crops. The usefulness of these techniques can be increased manifold by adopting natural resource management programmes and for sustainable agricultural productivity. The activities of AICRP on Dryland Agriculture particularly at Operational Research Project sites involve the adoption and evaluation of different soil and water conservation measures. For this, attempts have been made to conserve the natural resources like soil and water using earth moving machineries. The heavy machines have been used for the development of various water bodies viz., water harvesting tank, percolation tank, for drainage line treatment and even for the deep tillage operations to augment water requirement of different crops grown in the adopted villages. The results of these soil and water activities are found very promising and have been accepted and appreciated by one and all as it has increased the socio-economic condition of the farmers of the area. A brief account of all the land developmental activities is given in this book. Present book enlightens about such land developmental measures developed through heavy farm machineries which effectively conserved the natural resources in the region. We are sure that through these measures not only increase the productivity in the area but also bring new area under cultivation.

At last, we are sure that this book will be proven as a powerful medium especially for the students, professionals, field workers and farmers.

D.H. Ranade

S. Mujalde

Indu Swarup

Akhilesh Singh

D.V. Bhagat

O.P. Girothia

Contents

Chapter 1

Natural Resource Management through Land Developmental Process with Heavy Farm Machinery

Managing natural resources is a key factor for productivity in any of the watersheds. As per estimation, the per capita land availability is decreasing regularly due to development of wastelands besides other factors *viz.*, urbanization, increasing population *etc.* At the same time, uncontrolled movement of runoff water due to restricted drainage and/or obstructions in drainage line enhances the process of gully widening and deepening. The irregular shaped gully fragments the cultivable land reducing its area.

The productivity of the watershed can be increased by drainage line treatment, straightening of natural gullies (*nallah*) and breaking the continuity of sloppy land and converting it into terraced land. This will not only reduce the chances of overflowing of *nallah* but will also reduce the occurrence of sheet, rill and gully erosion due to reduced erosive velocity of runoff water. Similarly, it would be beneficial in terms of groundwater recharge if the fields near the ridgeline can be provided with diversion bund for regulating and controlling the entry of runoff water into the cultivable fields. Provision of diversion bund will not only act as barrier but will also retain runoff water on the upstream side. Wherever possible, the upstream portion may be converted into water bodies so that the stored water can seep through lower layers and can add to the nearby groundwater sources.

Availability of the irrigation water during the stress period is also a key factor responsible for increasing the productivity in the area. However, the water may not be available in plenty or even in small amount unless it is trapped during the excess period at suitable location. The runoff in a very large amount moves out of the fields, area, and watershed despite making all provision to reduce it using various soil and water conservation structures/ measures. This is mainly due to high intensity and amount during monsoon period and low infiltration rate of the soil, lesser soil depth *etc.* Thus it is always advisable to make suitable provision to store the runoff water in a storage area mainly in the discharge zone of the watershed or even in the lower portion near the field boundaries of a field. This will not only help in retaining the water in the area for a longer time to move it to go down deep in the soil to recharge the groundwater but also will provide sufficient water to use during stress period or even for irrigating the subsequent rabi crops,

However, the impact of these conservation measures varies with the locations and their effectiveness cannot be generalized. Thus, the advantage of providing various soil and water conservation measures should be assessed for the individual watershed.

Keeping these ideas in view, various studies have been during 1999 -2012 carried out in different watersheds of the region namely Baroli, Panod, Dakachya, Jaitpura, Gaddukhedi *etc.*, watersheds under **Operational Research Project for Dryland Agriculture, Indore** by undertaking various soil and water conservation activities through land developmental activities.

Problems of These Areas

The movement of uncontrolled runoff water develops gullies in the farmer's field. This process remains continue for the years together and increases the gully head advancement phenomenon. This not only increases the deepening and widening process of gully but also converts the cultivated fields into wasteland. Though it is difficult to reclaim the gullied portion, however, the straightening of gully may provide a passage for safe disposal of runoff and may provide a chance to utilize otherwise waste land. Since so far no such programme has been implemented to achieve such objectives, the present area was selected for the demonstration of such technology.

The runoff potential of the soils of the area is very high due to their low infiltration rate. The uncontrolled runoff enters into the farmer's fields with higher velocity particularly in the plots adjacent to ridgeline. This not only causes the severe sheet erosion but also develops gullies in the fields. The safe disposal/storage of this runoff would be beneficial. Since the soil of the area is underlain by fragmented basalt, the storage of surface water for a longer time would not be possible. However, construction of percolation tank will allow the downward movement of runoff water. Even the storage of runoff water in the farmers field with deep soil is not common in the region as in this case excavated type tanks are needed which are costly. At the same time, it is required that the waste land/uncultivated portion of the farmers be utilized for runoff water storage. Similarly, provision of diversion bund would divert the excess water away from the cultivated fields. The storage or disposal at reduced velocity of the runoff water would reduce the process of erosion and may increase the groundwater storage.

Tackling the Problems in Village Baroli (ORP Area)

Various activities pertaining to soil and water conservation *viz.*, straightening of gullies and utilization of wasteland, construction of percolation tank and drainage-line treatment were undertaken in the farmer fields ensuring their participation. The impact of such developmental activities is discussed in the foregoing paragraphs.

Straightening of Gully and Utilization of Wasteland

A reclamation process was started with the help of earth moving machinery (bull dozer). In this activity, a serpentine shaped gully with varying depth and width, which fragmented the cultivable field, was shaped into a straight-line gully keeping the width and depth almost uniform. While carrying out this operation, it was ensured that upper fertile topsoil of the newly located gully portion was first scratched and piled at a location. The excavated earth while straightening the gully was filled into the earlier serpentine shaped gully to convert all fragmented portions to one unit. After completion of filling work, the piled fertile earth was spread over the reclaimed/refilled portion to maintain its fertility. This whole process reduced the total length of gully by 20 m and the total cost incurred for bulldozing work was only Rs. 10,000/-. This not only allowed safe passage of runoff water but also helped the farmer to utilize extra area (0.18 ha) for cultivating different crops. The comparative details of this experiment are as under in the subsequent year, the work of gully

Straightening of Gully at Baroli Village.

Sl.No.	Particulars	Before Commencement of Work	After Completion of Work
1.	Length of gully (m)	110	90.0
2.	Depth (m)	Variable – 0.5 to 1.5	1.0
3.	Width of gully	Variable	4.0
4.	Cultivable land (ha)	0	0.18

Straightening continued so that the drainage line treatment in the upper zone is completed. In this phase, a total length of 400 m gully was straightened and the property bunds of adjoining fields were reshaped so that the runoff does not enter in the gully. Due to the bulldozing work, an additional area of 0.20 ha was brought under cultivation at a total cost of Rs. 5600/-, which was lying unused for the requirement of soil and water conservation measures, mainly for the safe disposal of runoff.

Sl.No.	Particulars	Before Commencement of Work	After Completion of Work
1.	Length of gully (m)	70	40.0
2.	Depth (m)	Variable – 0.5 to 1.5	1.0
3.	Width of gully (m)	Variable	4.0
4.	Cultivable land (ha)	0	0.2

Bringing New Area under Cultivation after Gully Straightening in Baroli

Further work on gully straightening and bringing additional area under cultivation was undertaken in the year 2002. In this particular portion of the area, a gully of serpentine shape was reshaped in a straight line as far as possible. Since the gullied portion was low lying due to severe erosion, a sort of terracing was provided in this piece of land with little leveling. Only Rs 25000/- was paid for the bulldozing work, which brought a huge area of 0.50 ha under new cultivation.

Sl.No.	Particulars	Before Commencement of Work	After Completion of Work
1.	Length of gully (m)	130.0	72.0
2.	Depth (m)	Variable – 0.5 to 3.5	1.5
3.	Width of gully (m)	Variable	3.0
4.	Cultivable land (ha)	0	0.5

Thus using earth moving machinery, uncultivated land of 0.88 ha was brought under cultivation which was lying barren/fallow for more than 20 years by investing only 40,000/- for gully straightening, leveling, terracing and removal of drainage line obstructions. The process has increased not only the productivity in the area but also increased the cost of land by 5 lakhs.

Percolation Tank for Groundwater Recharge

The details of the percolation tank constructed in the recharge zone in the watershed area are given below:

Storage capacity (ha-m)	:	0.50
Length of Bund (m)	:	190
Maximum depth (m)	:	2
Catchment area (ha)	:	8
Tank area (ha)	:	0.76
Cost of construction (Rs.)	:	40,000/-

The construction of percolation tank resulted in the storage of runoff water in the tank area in the very first year of its construction. Despite many dry spells in year 1999, all the runoff water retained in the tank area got percolated and increased the groundwater recharge. This was evident from the presence of higher water level.

Throughout the year in the nearby open well, which provided extra five irrigations to the *rabi* crops in the year 1999 -2000.

Similarly in the year 2000, project area experienced only 436 mm rainfall and even in this dry year, the open well retained 6m water in the month of January, 2001. Though no crop was irrigated by utilizing the well water during the year, the livestock available with the farmers was provided drinking water through the well. Further, this was also to be noted that all the other open wells in the area remained dry throughout the year.

In the year 2001, the project area again experienced only 741 mm rainfall against the average rainfall of 968 mm. Even in this dry year, the open well still recharged immediately after first heavy shower and retained reasonable amount of water for providing pre-sowing irrigation to *rabi* crops. It also provided again the drinking water to the livestock being maintained by the farmer. Once again, it is to be noted that all the other open wells in the area remained dry throughout the year.

Thus, the provision of percolation tank near the ridgeline not only retained the runoff to recharge the nearby well but also saved the adjoining field from sheet erosion and further development of gullied portion.

Drainage Line Treatment

As already indicated, ridgeline of this watershed is very prominent; therefore, the runoff potential is very high in this region. The uncontrolled runoff developed a seasonal gully in this watershed. The depth and width of this gully was almost non-uniform throughout the whole gully length. The maximum and minimum width being 20 m and 2 m with depth in the range of 0.5 to 2.5 m and change in the runoff velocity due to gully cross section erodes farmers' field boundary. For this purpose, bulldozer was used to

make this *nallah* in a regular shape as far as possible. The de-silting of gully was also carried out to make its depth uniform. The de-silted material was put up on the field boundary to strengthen both banks of the gully. This process not only reduced the further cutting of *nallah* in irregular shape but also reduced the process of sheet erosion from the adjoining fields. In all, about 250 m length of *nallah* was treated and the cost of bulldozing work was only Rs. 10000/-.

The developmental activities undertaken in the watershed area using heavy machineries could bring fruitful results by bringing additional area under cultivation. The drainage line treatment, land leveling and terracing also reduce the chance of further development of gullies and runoff induced soil erosion.

Drainage Line Treatment at Baroli Village

Development of Water Bodies in Village Barlai

As per the requirement of the area due to erratic monsoon condition, the suitable sites were selected and three water harvesting tanks were created in the farmers fields so that the stored water could be utilized for the irrigation of crops during moisture stress period. The participation of the farmers was fully ensured for this and approximately 50 per cent of the cost of tank was borne by the farmers in terms of kind or labourers/use of tractors, trolley and implements like MB plough, leveler *etc.* The land for the development of the tank was donated or sacrificed by the farmers either from the wasteland/unused land or from the part of cultivated fields. The stored water would be utilized during the stress period or even for the paleva while sowing wheat, gram *etc.*

The details of the tanks are given below:

Particular	*Deepak Tank*	*Patel Tank*	*Bhura Tank*
Type of the tank	Excavated cum Embankment	Excavated	Excavated cum Embankment
Outlet	Pipe	Boulder spillway	Pipe
Length of the tank (m)	95	55	90
Width of the tank (m)	43	40	32
Max. depth (m)	2.35	2.75	2.75
Storage capacity (Cu.m)	3580	2670	2538
Cost incurred from project for bulldozing work (Rs)	48400/-	70000/-	90000/-
Farmers share (Rs)	40000/-	70000/-	15000/-
Total cost (Rs)	98400/-	1,40,000/-	105000/-
Estimated cost through manual labour (Rs)	200000/-	250000/-	200000/-

Bringing New Area under Cultivation through Excavated Soil

As already pointed out the storage of runoff water is essential for its utilization for saving the crops during stress period and for providing pre sowing irrigation and post sowing irrigation during *rabi* season. The selection of the proper site for the construction of water harvesting tank is dependent of various factor. But in the leveled field, the construction of tank is carried out by excavating the soil to provide the impounding of water. The excavated portion has the fertile soil essentially at the top layer which can be transported to:

Three Tanks in Village Barlai

1. To spread in the undulating portions of the field so that the slope of the field is reduced for increasing the time of concentration and to reduce the erosive velocity of the runoff water and to avoid the chances of sheet erosion;
2. To fill the depressed portion in the fields for easier cultivation operations;
3. To spread the fertile soil on the waste land area or uncultivated area to make the same area suitable for the cultivation.

In the present case in the village Barlai, the excavated soil of the new tank (Deepak tank) area was lifted through tractor trolley by spending only Rs. 5000/- in labour and fuel and was spread over a fallow undulating area (120m x 6m) in the month of May-June 2003 which was lying unused for the decades due to undulating topography. The same area was brought under cultivation immediately during the *Kharif* season. The portion has been

Deepak Tank

Patel Tank

Bhura Tank

planted with the soybean + sorghum crop which produced about 10q/ha soybean + 30q/ha sorghum. If the market value of such portion is calculated even at a very conservative rate, it was observed that the farmer is benefited by Rs. 100000. Again this new area is planted with gram/wheat during *Rabi.*

Removal of Weeds from the Cultivated Fields

It is a well known fact that the development of the weeds in the cultivated fields has a very adverse effect on the crop growth and ultimately the crop productivity. To get rid of such weeds number of weedicides are spread in the cropped area which are very costly, time consuming and labour intensive. If the fields are made weed free through mechanical means, the overall cultivation cost can be reduced and thus the farmers can make a profit out of his farming system.

Brining New Area under Cultivation by Spreading Excavated Soil

Sometimes, it is observed that the lateral roots of the trees grown/ planted on the field boundaries enter into the fields' laterally even up to 100 m inside and interfere with the cultivation operations. These lateral roots may develop suckers during the cropping period and thus become competitor with the crop. This reduces the overall productivity of the land.

Such fields in the village Barlai were identified. About 12 ha of such fields were then ploughed using a M.B. Plough which is operated by the crawler tractor and may turn the soil up to 45 cm. At this depth, the lateral roots were quite visible and were then destroyed due to deep ploughing operation. The destroyed roots were collected and removed from the fields. The removal of such weeds in large quantity thus allowed production of good crop despite the formation of dead furrows and slight disturbance in the field leveling. However, chances of field disturbance can be avoided if reversible plough is used while deep ploughing.

Removal of Weeds with Deep Tillage in Barlai

Thus various soil and water conservation measures have been created in different watersheds of the region utilizing the heavy machines for such work. The creation of water bodies, straightening of the *Nala*, drainage line treatment brought about visible changes and helped in managing the natural resources in the region. The use of heavy machines not only completed work in time but also saved the expenditure which would have been otherwise

almost double if the manual force was utilized. Therefore it can be pointed out that besides man, money, materials, machines also play a very significant role in natural resource conservation programme.

Chapter 2

Innovative Approaches for Managing Harvested Irrigation Water for *Rabi* Crops in *Malwa* Region

The major constraint in growing a second crop in many areas particularly in *Malwa* region is the lack of suitable irrigation water. Therefore, creating water storage tanks in the cultivated field itself for storing the excess runoff (common feature in this region due to low infiltration rate of the soil and high intense rainfall), is one of the best options available with the farmers. In *Malwa* region, the topography is such that two types of the tanks can be constructed in the farmers fields:

1. Excavated tanks suited for flat topography and
2. Excavated cum embankment type tanks particularly in degraded or gullied portion.

In the first case, provision of inlet in structural form is not required as flow enters through entire width of the tank. The scientists of Operational Research Project for Dryland Agriculture located at College of Agriculture, RVSKVV, campus Indore have exploited various such sites and provided the technical guidance, improved the technology and helped many farmers for storing excess rainwater in dugout farm ponds for utilization during the dry spells of monsoon season or for raising a subsequent crop and further suggested suitable cropping pattern for improving the productivity. Now it is a well established fact that activity of excavation of farm ponds on the individual farmer's field was extremely useful in turning the mono-cropped area into multi-cropped/crop diversified area and improving the farmer's income and, thus, his/her standard of living.

Beside the facts, many interesting experiences have been gathered by the ORP team particularly for the effective utilization of the stored and available water in the ponds.

Generally in this region, the tube wells provide groundwater with reduced discharge after the termination of monsoon season as groundwater depletes at faster rate due to overexploitation. Thus generally very few tube wells operate throughout the year otherwise most of the tube wells go dry or supply water only for 2-3 hours that too with very reduced quantity. At the same time electric power supply is a deterrent in providing continuous irrigation to field's crops. Sometimes the supply of tube well is so less that direct irrigation is not possible as water advancement process does not take place particularly in medium and deep clayey soils due to cracking pattern and irrigation process is not completed despite pump running for several hours.

Therefore the well or tube well water is collected in the tanks for few days and then the stored water is used for irrigating the nearby fields using a high discharge diesel/electric pump so that a larger area can be irrigated at a time. Similarly since the percolation rates of these tanks are very meager, the farmers are collecting tube well water in these tanks for its utilization for raising summer/vegetable crops. Thus the farmers rely more on stored water in the tanks than the tube well water as most of the tube wells dry up suddenly/abruptly due to groundwater depletion.

In this region, due to inherent properties of *Vertisols*, severe cracks are observed on the surface in general after the termination of monsoon and harvesting of soybean crop and particularly at the time of sowing of *rabi* crops. For this purpose, farmers either provide pre sowing border irrigation or border irrigation after dry sowing (if possible due to residual moisture/late rains). Because of this, high capacity pump (3-5 HP) is required so that the water front moves rapidly and covers the whole area, otherwise the cracks will not allow the water to move fast and irrigate the entire field. In a study it has been reported that for irrigation purpose, the farmers utilized 5 HP diesel pump or electric pump having discharge rate of 10 lps. Since the rainfall pattern was very erratic during 2005 with only 28 rainy days, it was observed that for pre-sowing irrigation, the depth of irrigation varied from 16 to 21cm as water was going deep into the soil through cracks without irrigating/moving on the field surface). Similarly during 2005, for irrigating one-hectare area, 70-75 pumping hours were required. However during 2006, the depth of irrigation water required for pre-sowing irrigation was limited only 10 to 11 cm with only 30-32 pumping hours as the area received good and well distributed rainfall with 59 rainy days.

Many a times, it has been observed that for the utilization of the water, farmers face typical problems. Utilizing 5 HP diesel pump all the time is quite expensive, and also not easy to transport it to tank site and thus farmers normally do not prefer it and use electric pump of 5 HP instead. However, the irregular supply of the electric power is a major hindrance for the timely pumping of the stored water using electric pump. At the same time, farmers are required to be at the tank site while pumping due to possibility of theft of the pumping unit. Therefore, pumping has to be carried out as early as possible so that the pumping unit is transported back to home safely.

After the construction of tank in the field of Mr. Pradip Shidye, village Barlai Jagir, Tehsil Sanver, District Indore, the following unique and innovative idea came into the minds of the group of farmers and Scientists that why not such a unit is designed so that the entire irrigation operation is completed in one go using all types of pumps viz, electric, diesel and tractor PTO driven so that the pumping is not to be stopped for the want of electricity. Such a devise is ultimately designed and used successfully. The detail of such unit is given below:

Portable Electric-cum-PTO Driven Pumping System

Mr. Pradeep Shridhar Shidhye, Village and Post: Barlai Jagir Tehsil Sanver, district Indore (M.P.) is a progressive farmer. The electric power supply in his area too is very erratic and irregular. Thus dependency on electricity for irrigating the fields using harvested water from nearby tank will not only delay the *rabi* sowing but also allow the stored water to get percolated without irrigating the crops. Therefore, it was thought that provision is made in such a way, that water can be lifted for irrigating the crops easily even in the absence of electricity. The system should also be easily taken to field and the farmer is not needed to drag the heavy diesel

pump to the fields. After the work, again it carried along with tractor to the farm shed.

Mr. Padeep utilized a frame of old cultivator and put a pump on one corner of the frame. On the other end, an electric motor was fitted. Similarly in the middle of the frame, a pulley with suitable gears was also fitted. Provision was made to operate the pulley through PTO of the tractor and simultaneously with electric motor. It was so fitted that through pulley, pump can be operated with V belts either by using electric motor or by PTO of the tractor. Thus, he could irrigate his fields by lifting harvested water using electric motor when electricity was available otherwise through PTO of his tractor without waiting and wasting the time. The system is easily taken to the fields without any problem and the farmer is not required to carry otherwise heavy diesel pumps to the fields which is very tedious work.

Utility of the System

1. Can be operated for irrigating fields even if electricity is not available.
2. Old frame of cultivator can be used.
3. Tractor is utilized even during non farming time.
4. Can be transported to the tank site very easily.
5. Can be fitted by the farmer himself without any technical support.
6. Best alternative to electric pump with higher HP power.
7. Better than portable and pumps with lower discharge rate.

Many farmers of the region are visiting his farm and trying to use such useful innovative idea in their fields. 12-13 farmers of the region having water harvesting tanks are using such system for irrigating about 50-60 ha area during *rabi* season. No, commercial unit is required as farmers have to use only their available resources, implements, pumps and motors just to be fitted on cultivator frames and to be driven by PTO with tractors.

The innovative idea came only after the construction of water harvesting tank in his field through one ICAR funded project by ORP team. The availability of the water in the tank gave him confidence and only then this innovative idea came in his mind during the discussion at the site. Similarly, the availability of water encouraged him for goat rearing.

Chapter 3

Efficient Water Irrigation and Saving Device for Border Irrigation of *Rabi* Crops

In most of villages, due to peak hour requirement of electric power, it is normally provided in most of the villages during night hours particularly during the irrigation period of *rabi* crops. In that case, it is really impossible, too cold and very tedious to be in the fields' whole night for irrigating the crops. Therefore, the farmers normally put the pump on till the electric power is available and allow the water at higher pumping discharge rate to move in the fields on their own through pipes without attending it all the time. This system not only over irrigates the fields but also wastes the precious water and allows it to move out of the fields without covering the maximum possible fields under irrigation. Therefore, it was thought that if the borders are irrigated through perforated pipe so that the discharge rate is reduced and water movement in the border is controlled (with reduced velocity) and water does not move out of the fields and water is not wasted. At the same, time if the pipe is rested on adjustable stands/platform, the height of the pipe is adjusted in such a way that the three borders receive water at uniform discharge. Depending upon the discharge rate of the pump, the coverage of the borders can be increased.

Mr. Laxman Singh Chouhan, Village and Post: Shriram Panod, Tehsil Indore district Indore (M.P.) has made arrangement for irrigating the

24/12/2007

24/12/2007

24/12/2007

24/12/2007

24/12/2007

24/12/2007

24/12/2007

chickpea or wheat grown in borders. Perforations have been made in the PVC pipes attached to rubber tubings at suitable spacing as per the border width. The pipe is then rested on three adjustable platforms or stands so that the pipe covers three borders at a time with perforations for irrigation. Now, if pump is operated, these perforations allow the water at a uniform discharge to fall in the border width only. If the fields are undulating then pipe height is adjusted to irrigate the borders with uniform discharge rate. The advantage of this system is that all the three borders are irrigated simultaneously with a 20ft PVC pipe. The coverage of borders can be increased as per the discharge rate of the pump and it is so adjusted that the system works for more then 7-8 hours without attending and keeping vigil whole night. Using such a system, five pipes of 20 ft can be operated at a time and borders in 30 m length are irrigated.

Utility

1. More area is covered with same quantity of water.
2. Avoid over irrigation of the fields.
3. Can be run whole night without attending the system.
4. The discharge rate can be adjusted in all the three borders with the help of adjustable stand.
5. The borders are irrigated simultaneously.
6. Can be made in local workshop.
7. Very useful under limited power supply at night.
8. Dependency on the labour can be reduced.

Many farmers of the region visiting his farm and trying to use such useful innovative idea in their fields.

The innovative idea came only after the construction of water harvesting tank in his field through one ICAR funded project by ORP team. The availability of the water in the tank gave him confidence and only then this innovative idea came in his mind during the discussion at the site.

Chapter 4

Use of Water Harvesting Tanks as Storage Structure

The various tanks constructed under the scheme are being utilized by the farmers in many ways. The runoff water collected during monsoon season not only provided life saving irrigation to *kharif* crops during dry spells but also ensured planting of *rabi* crops during water scared season and scanty rainfall year and even during normal rainfall years. The pre-sowing irrigation before *rabi* planting allowed farmers to harvest higher grain yields even other adjoining field remain uncultivated for the want of soil moisture profile. Similarly in few cases, the stored water is being utilized for fish rearing too. However, not much encouraging results have been generated on this particular aspect because of obvious reasons during the study period. On the other hand, even during good rainfall year the tanks constructed under the schemes are benefiting the farmers in many ways. Generally in this region, the tube wells provide groundwater with reduced discharge after the termination of monsoon season as groundwater depletes at faster rate due to overexploitation. Thus generally very few tube wells operate throughout the year otherwise most of the tube wells go dry or supply water only for 2-3 hours that too with very reduced quantity. At the same time electric power supply is a deterrent in providing continuous irrigation to field's crops. Sometimes the supply of tube well is so less that direct irrigation is not possible as water advancement process does not take place particularly in medium and deep clayey soils due to cracking pattern

Tank Refilling with Groundwater in November 2008.

Tank in July 2009.

Tank on 20.07.2010.

Refilling with Groundwater 12.11.2010.

11/15/2009

11/07/2008

11/02/2010

11/02/2010

17/11/2006

11/07/2008

10/19/2010

10/19/2010

Typical Pumping Unit Arrangements for Irrigating *Rabi* Crops (2010-2011).

and irrigation process is not completed despite pump running for several hours.

Therefore the well or tube well water is collected in the tanks for few days and then the stored water is used for irrigating the nearby fields using a high discharge diesel/electric pump so that a larger area can be irrigated at a time. For this purpose, arrangements are to be made operating diesel and electric pump as and when required. Similarly since the percolation rates of these tanks are very meager, the farmers are collecting tube well water in these tanks for its utilization for raising summer/vegetable crops. Thus the farmers rely more on stored water in the tanks than the tube well water as most of the tube wells dry up suddenly/abruptly due to groundwater depletion. This practice has been adopted by many farmers in the region to tackle the duel problems of reduced discharge through tube-well and erratic electricity supply in the villages. An example of the water harvesting tank constructed in the field of Mr. Bhagwat Singh Chouhan, village Panod is given in pictorial form.

Chapter 5

A Low Cost Sunken Structures for Increasing Water Availability and Agricultural Production in Tribal Areas of Madhya Pradesh

Through Operational Research Project (ORP) for Dryland Agriculture, College of Agriculture, RVSKVV, Indore, the team members are involved in various activities including technical guidance to other allied institutes, agencies for the upscaling of various technologies for increasing overall agricultural productivity. In few the watershed development programmes operated in tribal areas of Dhar, Jhabua and Ratlam in collaboration with Department of Agriculture and Farmers welfare, MP government during (2005-2013), the team members of ORP were actively involved as resource personnel for providing technical guidance. The experiences shared while working in the tribal districts by them are being produced here.

During Participatory resource appraisal (PRA), it was observed that villagers were adopting an indigenous technology of excavating in stream bed to increase percolation of water for a long time (locally known as Kundiya/DOH) which is certainly a low cost water harvesting technique. Therefore, as per the policy, it was decided to popularize this indigenous technology through demonstration, improvement through perfect blending of local and technical scientific knowledge and to provide technical guidance for its propagation.

A significant portion of the project area is barren and used for open grazing. The sunken water harvesting technology was initially demonstrated

Indigenous Technology Knowledge (*Kundiya*).

Degraded Land with Exposed Rocks.

Eroded Fields with Poor Fertility.

in few villages as an alternative to the high cost heading type approach to water harvesting being implemented on a large scale by main line programmes.

The topography of these areas is undulating with large patches of non-arable lands and small patches of arable lands in valley portion. The entire area is devoid of vegetation for 6 months of the year. Non-arable lands are highly degraded with large portions of exposed rock visible. Small to medium gullies are observed up to bedrock level.

Specification of Sunken Structure

Definition of Technology

Sunken structure is constructed by excavation in stream bed with a view to provide temporary storage of runoff, increased percolation for recharge of groundwater, subsequently tapped through shallow wells or *odees*.

Description

The *doh* comprises an excavated portion (rectangle shaped) of 1 to 1.5 m depth and variable length and width as per stream section. The excavated material is deposited along the side banks with a view to prevent siltation from surrounding areas. An upstream slope of 1:6 and downstream slope of

1:8 is provided to facilitate movement of silt along with flow thus reducing siltation. A silt trap of loose boulders is provided on upstream side. The structures are generally constructed in series to trap maximum runoff, however bends in the stream should be avoided as these are susceptible to bank erosion. The technology is used in conjunction with shallow wells adjoining the stream (*odee*), which enable farmers to harvest the increased sub-surface flow for irrigation.

Doh is a structural measure constructed in a defined gully portion with a view to conserve water for better runoff management along with reduction in land degradation of adjoining fields. The problem of water conservation and gully erosion is generally addressed by *DOH*. It is a means of controlling concentrated runoff to increase its infiltration and to harvest the runoff water. Thus it is a technology intended to recharge the wells located downstream or in the adjoining fields as a result of increase in groundwater levels.

Doh technology is relatively new which is a modification of the traditional *Doh/Kundiya* observed in river beds. *Kundiya/Doh* is a naturally occurring depression in river/*nallah*/gully bed where water is collected and then used by villagers and cattle. However, due to advent of piped water schemes linked to bore-wells, the traditional technology is now being neglected.

Idea of Propagating Sunken Structure is to Discourage High Cost Heading Type Approach like Masonary Structures/Stop Dams.

Purpose

- ☆ Sub-surface storage of run-off,
- ☆ Groundwater recharge through increased percolation,
- ☆ Flood control through series of such structures.
- ☆ Low cost alternatives for poorer communities.
- ☆ To discourage/alternative to reinforced masonary structure.

Site Selection

Generally the site selection for *Doh* is carried after consultation with village community and keeping in mind the requirement of beneficiaries. However for the construction of *Doh* following points should be considered for selection of proper site to get the maximum benefit from this technology.

- ☆ Gradually undulating topography,
- ☆ Well defined gully/stream sections,
- ☆ Weathered bed strata which can be easily excavated and at the same time increase the percolation rate.
- ☆ The stream should have a straight reach on both the sides: u/s as well as d/s.
- ☆ Gully bed slope should not be more than 3 per cent.

- ✩ Gully should have stable banks or banks are to be protested using boulders/vegetative measures.
- ✩ *Doh* should not in any way lead to water spreads into the nearby agricultural fields.

Further, in any case the *Doh* should not be constructed:

- ✩ In areas where the foundation strata allows water to emerge immediately from the downstream side of *Doh*.
- ✩ Sites with fracture or dyke, which may convey out water from the storage area.

Thus for the construction of *Doh*, the most important natural conditions would be gradually undulating topography, well defined gully/stream section and weathered bed strata. Since streams are regarded as common property, ownership/land use right less affects the location of *Doh*, however use of water is affected, since, nearby farmers have easier access to the water. At the same time, the de-silting process of *DOH*, may increase the productivity in the adjoining fields if the excavated soil is spread in these fields.

However after or before the construction of *DOH*, it is required to treat the catchment area with gully plugs and other measures so that the silting is reduced in the *DOH* area.

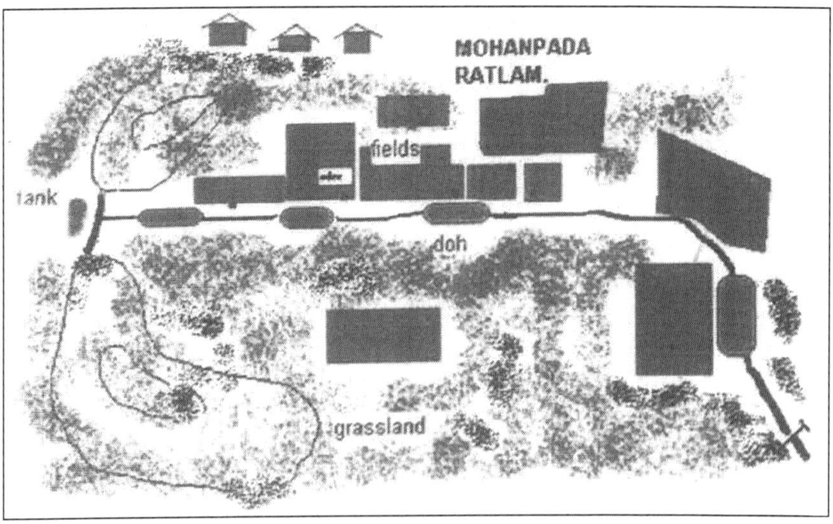

Location of *DOH* in Gully.

Design of *DOH*

Generally *DOH* is a rectangular shaped structure but the final shape and upstream and downstream slopes are dependent of the specific site and bed strata.

Technical Drawing

Section of Typical *DOH* Structure

(A) Original *Nala*

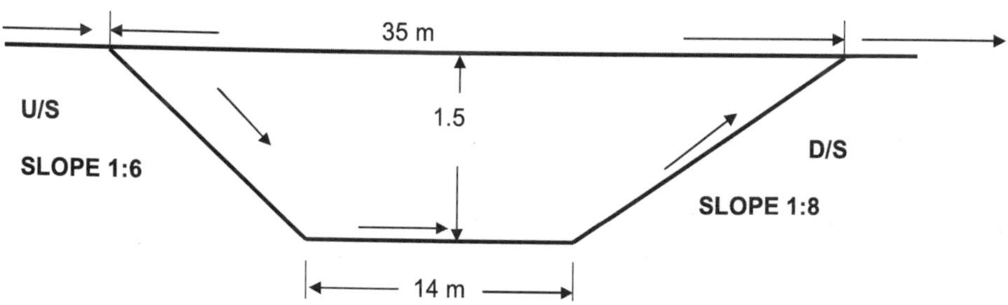

(B) After Excavation

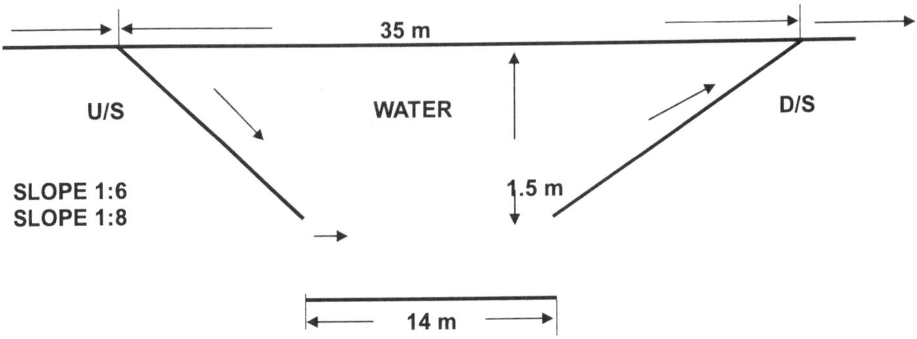

(C) After Rains

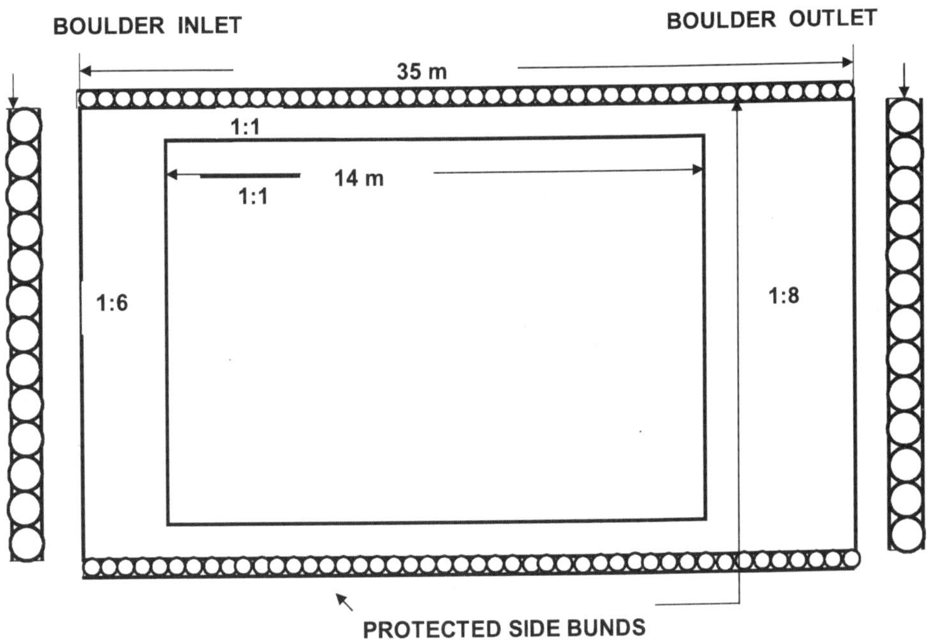

Plant of *DOH* Structure.

Specification of *DOH*

- ✫ Vertical interval (between two DOH) – 1.5 m
- ✫ Horizontal interval (in project condition) – 50 m
- ✫ Length – 35 m
- ✫ Width – 8 m
- ✫ Depth – 1.5 m
- ✫ Downstream slope – 1:8

- ☆ Upstream slope – 1:6
- ☆ Height of side bund/Bank – 0.5 – 1 m
- ☆ Width of side bund – 1-2 m
- ☆ Length of each side bund – 30-40 m

For the construction of *Doh*, weathered section in the gully is excavated manually. But the hard strata (hard rock beds) may be excavated by

***Doh* with Boulder inlet and Outlet.**

Side Protection of *Doh* with Boulders and Vegetative Measures.

machines. Since the excavation of *Doh* can be carried out during dry season or after kharif season, this labour intensive work provides employment to otherwise migrating persons located in the watershed area. The supportive technologies for *Doh,* which are desirable but not essential, are stone line/ bund across the major slope/contour, contour cultivation and boulder checks on rills and gullies in upper catchment.

Location of *ODEE*

Generally *Odee* (shallow wells) are constructed in the adjoining fields to trap the seepage water laterally moving through weathered strata or basaltic murrum due to storage of water in *Doh* area. However, few cases of construction of *Odees* in the *Doh* area have also been observed. However, in

***Odee* in *Doh* Area and Use of Diesel Pump for Irrigating Crops.**

***Odee* Outside *Doh* Area.**

the later cases, the storage capacity of *Odee* is affected due to silt deposition. In this case, regular desilting of *Doh* and *Odee* is required to maintain their storage capacity as well to increase the number of silt traps to minimize siltation.

Due to increase in groundwater level and availability of water for longer duration, the farmers are now using diesel operated pump and even manually operated low hand pump for irrigating the cropped area.

Benefits or Advantages of *Doh* Technology

In these regions, during monsoon (*kharif*) season the main crop is cotton + maize. In addition, minor pulses are taken. If water is available for irrigation, this is followed by wheat + gram in winter. Major area is fallow in winter except for the cotton crop which is of longer duration. No cultivation was observed in summer. In the technology area, as a result of increased water availability for irrigation, farmers are now cultivating vegetables (chilli, tomato, Okra, gourds) and fruit crops like mango, Papaya, grafted Ber *etc.*

In a study, it has been observed that due to construction of *DOH* and resultant increase in groundwater availability, the production of Cotton and maize has increased by almost 300 per cent and 400 per cent respectively. The details are depicted in the following table.

Plantation of Fruit Crops Due to Increased Availability of Water.

With or Without	*Main Plant Type*	*Production in t/ha/Season*
Without SWC	Cotton	0.25
With SWC	Cotton/Maize	0.70
Without SWC	Maize	0.50
With SWC	Maize	2.00

Thus the benefits of *Doh* technology as observed in project area are:

Production and Socio-economic Benefits

- ☆ Crop yields have doubled and high value crops are being introduced/taken up in the project area.
- ☆ Farm income has increased.
- ☆ Community institution strengthening has been observed due to involvement of village watershed development committee and users group.

Ecological Benefits

- ☆ Increase in vegetation for longer duration
- ☆ Enhancement in soil moisture due to increased sub-surface seepage
- ☆ Increased runoff water management
- ☆ Decrease in soil loss due to silt trapped in gully plugs
- ☆ Increase in water levels in wells
- ☆ Increase in soil fertility due to spreading of desilted soils in the adjoining fields.

Other Benefits

- ☆ Drinking water availability has increased, as surface water is available for longer duration.
- ☆ Increase in water levels in well and hand pumps

Off Site Benefits

- ☆ Reduction in downstream flooding due to reduction in velocity and amount of runoff
- ☆ Reduced downstream siltation due to silt trapping in gully plugs,
- ☆ Even cases of increased stream flow in dry season because of increased subsurface seepage.

Constraints during Construction and Establishment

- ☆ Technical: Hard strata may require use of heavy machine/ Dynamite,

- Social: Implementation of decision not to lift water directly from *DOH* is difficult.
- Maintenance: Can only be done after harvest of the main *kharif* crop (November/December).

Possible Improvements

- Deepening of *Doh,*
- Stone pitching of Banks.

Maintenance

As already indicated that the supportive technologies for *Doh,* which are desirable but not essential, are stone line/bund across the major slope/contour, contour cultivation and boulder checks on rills and gullies in upper catchment. However, if these supportive technologies are not properly supplement *Doh* technique, the silting of *Doh* is inevitable. Therefore, after or while, construction of *Doh,* before and after maintenance schedule should be followed to sustain *Doh* technology.

Silting and Reduction in Storage Capacity of *Doh.*

Advantage of Sunken Structure Over Surface/Heading Type Structure

Sl.No.	*Sunken Structure*	*Other Structures (Check Dam, Cemented Structure etc.)*
1	Occupies less area and much less submergence of land, small exposed water surface.	Occupies more area, more submergence of land, more exposed water surfaces.
2	Deep water column, less evaporation and greater percolation into surrounding areas.	Water column is less. Exposed area is more and more evaporation and less recharge.
3	Sub surface flow is more.	Sub surface flow is less.
4	Generates more employment for the families of watershed area.	Less employment opportunities, since it requires skilled labour.
5	Major portion of money goes to local people.	Major portion goes to materials which is not available to the local people.
6	Easily manageable by local people.	Requires skilled labour and heavy investment.
7	It protects and increases life of the structures of the drainage line because of *in situ* conservation of soil.	Without ponds at the upper reaches danger of breaching and siltation of structures.
8	Benefits more people especially poor and marginal.	Benefits less number of people and merely rich people.
9	Equity in water distribution.	Unequal water distribution in the watershed.
10	Very effective microclimate management.	Little effect on the microclimate.
11	Energy efficient structure.	Loss of energy requires additional energy to bring backwater to upper reaches.

Chapter 6

Coordinated Approach for Developing Water Storage and Water Recharging Structure for Farm Income and Crop Productivity Enhancement

Besides many mandates, the Operational Research Project for the Dryland Agriculture, College of Agriculture, RVSKVV Campus Indore has been actively working in different watersheds of *Malwa* region to achieve a firsthand working experience in the development of area on watershed basis to serve as a model for extension agencies and to provide consultancy services to the extension agencies for transfer of dryland technologies.

In the year 2010, as per the requirement and ICAR directives, a new ORP village Gaddukhedi (22°48′N and 75°50′E, altitude 540 m) was finalized by considering various aspects and in consultation with administrative and Agriculture officials of Dewas district of Madhya Pradesh. After making several visits, a suitable ORP site is selected. Before formulating the technical programme, deciding research strategies, conservation approaches and technologies, it is desirable to collect basic information on various aspects. A detailed Participatory Rural Appraisal (PRA) was conducted to establish rapport with the village community as well as to identify and define problems for prioritization in this village itself by the ORP team, along with many officials of Department of Farmer's Welfare and Agricultural

Development, Dewas. The idea was to dovetail the activities in the village Gaddukhedi at least for some development work.

It was observed that the *Vertisols* of the village are low in Nitrogen content with medium P and high K level. The soil is very shallow near the ridge line and in the middle reaches; soils are medium to deep in nature. Further, it has been classified as *rainfed agro ecosystem and oil seed based production system*. The average rainfall is 1067 mm with average 45 rainy days. Thus, it offers potential (apart from the above identified need) for rainwater harvesting in the district. The runoff potential in this village is very high due to their low infiltration rate. The safe disposal/storage of this runoff would be beneficial. The storage of the runoff water would help the farmer to utilize it during the moisture stress period. Since the soil profile is underlain by basaltic *Murrum*, even the stored water would no longer be retained for longer time. But its percolation may enhance or recharge the underground tube well or open well available in the vicinity. Further, in most of the rainfed areas, it has been observed that in *kharif* season rainfall is adequate for agriculture; irrigation is required for *rabi* and summer season. Even in village Gaddukhedi, due to poor/non-availability of groundwater, lot of area could not be brought under cultivation during *rabi* season many a times in this village. Since runoff and associated soil losses thereby reduced fertility, non-availability of surface as well as sub-surface water, limiting moisture conditions has been the major problems of the region, it was thought that attempts should be made to water availability in the farmers fields so that besides *kharif* crops *i.e.*, mainly Soybean, Maize, other *rabi* crops like chickpea and wheat are grown without any fail at least in few area. Similarly, this would be an entry point activity for ORP so that enhanced water availability would ensure the cultivation of *rabi* crops for further help in achieving the other objectives *i.e.*, to evaluate the performance of each component of dryland technology under the farmers management conditions both singly and in combination and to provide feedback to the research stations for fine tuning sub-optima recommendations. However, in a watershed area, it is not possible to assess the direct or indirect advantage of this water harvesting cum recharge tank on many fields and farmers away from the tank area. Therefore, it was decided that the data are generated from a farmer who is directly drawing water from this tank area and from his open well located very nearby hardly 3 m and connected through tank with a pipe for recharging purpose. Thus, this study was carried out 2010 onward with an objective to generate data on water availability and to observe its effect on increase in *rabi* cropped area.

Water Recharging-cum-Harvesting Tank for Storing Runoff Water

It was, therefore, decided to increase the runoff water storage in the nearby existing natural gully or *Nala* by creating storage pockets located in the vicinity of selected farmer's field which is otherwise runs out of the field after creating sheet erosion losses and development of washes. A suitable site in the field was selected considering all the hydrological aspects in April 2010 during the PRA exercise. In this portion, technical guidance was provided to line department personnel and an excavated tank was developed using watershed funds available with line department and finally a suitable cemented structure in the form of outlet was constructed totally under the guidance of ORP team in June 2010. This tank of 1500 cu. m capacity was allowed to fill with the runoff water in the following monsoon season and as expected most of the water percolated and very limited water was available for the irrigation purpose.

Table 6.1: Details of Pre-sowing Irrigation through Tank Applied during *Rabi* Season

Particulars	*Year*				
	2009-10	*2010-11*	*2011-12*	*2012-13*	*2013-14*
Amount of Water used (m^3)	-	1500	3000	3000	3000
Area Irrigated (ha)	1.5 (using tubewell)	5 (conjunctive use)	10 (conjunctive use)	10 (conjunctive use)	10 (conjunctive use)

Further, it was then decided to make use available of the cemented out let waste weir during March 2011. For this purpose the existing *Nala* portion other than the tank area towards upstream was excavated using the back hoe loader machines so that the back water retained in the *Nala* and may aid to nearby open well. Participatory mode was adopted for this purpose and farmers was asked to bear the cost of excavation by providing his tractor and trolleys so that the excavated soil is spread in the undulating fields of the farmer's fields and the basaltic material is used for the construction of field bunds, Also the excavated basaltic material was used at the field bunds for providing waste weirs in the bund portion so that the silt free water from the cultivated fields enters into the *Nala*. This brought a sizable portion of the farmer's fields into leveled condition and also provided a safe disposal to runoff water. Because of increased capacity of the pond area due to excavation, full advantage of the waste weir was explored. An additional storage of 1500 cu.m was possible because of this activity during the monsoon period of 2011-12. Immediately after the harvest of soybean, the adjoining fields (10ha) were irrigated using the submersible pumps. The stored water was used for pre-sowing irrigation and subsequently for two more additional irrigations. The excavated portion also ensured groundwater recharge in the nearby open well and this ensured operation of tube well located in the open well for additionally for one more month than the others located in various fields.

Table 6.2: Change in *Rabi* Cropping Pattern due to Additional Availability of Water in the Tank and Groundwater

Crop	*Year-wise Area under Rabi Cultivation (ha)*				
	2009-10	*2010-11*	*2011-12*	*2012-13*	*2013-14*
Gram	7.5	5	2	2	2
Wheat	2	4.5	7	7	7
Potato	-	-	0.5	0.5	-
Vegetables	0.25	0.25	0.2	0.2	0.2
Sweet potato/groundnut		0.05	0.1	0.1	0.1
Sorghum fodder	0.2	0.2	0.2	0.2	0.7

From ORP side, the advantage of this outlet was explored and through this sunken pond, the nearby open well gets recharged automatically in the monsoon season of 2011-12. This helped the farmers to bring an appreciable amount of cultivated area under additional irrigation. Similarly, in the year 2012-13, the advantage of the increased storage capacity was very much visible as the open well located in the adjoining field remained fully recharged and allowed farmer to draw more water than other wells located in the same area but away from the influential zone of the water harvesting tank. Similarly, during 2013-14, the heavy runoff due to more rainfall resulted in the tank remained full to its capacity almost throughout the monsoon period. The adjoining wells and tube-wells got recharged and provided extra water available for irrigation. The farmer utilized the stored surface water by using floating submersible pump during the month of October and November to provide pre-sowing irrigation to adjoining fields and brought the fields under cultivation of gram and wheat. It is to emphasis that the site was such selected that despite erratic rainfall pattern, the tank area and extended portion remained full throughout the rainy seasons of 2010, 2011, 2012 and 2013 and provided surface irrigation water for pre sowing irrigation in 10 ha area. This has allowed him to bring additional 4 ha area under wheat (Now it is 7 ha wheat, 2 ha chickpea and 1 ha other crops). Otherwise he was somehow managing to grow, chickpea in 8 ha and wheat in 2 ha. This has allowed farmer to diversify his cropping pattern and he could grow potato, summer maize, vegetables and sweet potato for his own use and family which was otherwise not possible before.

Table 6.3: Average Increase in Yield of Chickpea and Wheat Due to Extra Tank and Groundwater

Crops	*With Irrigation (kg/ha)*	*Without Irrigation (kg/ha)*	*Increase In Yield (kg/ha)*
Chickpea	1600	900	700
Wheat	5600	3400	2200

It was observed and monitored that even in the month of October throughout all these years, the open well of the farmer was full when there was no water in the all the three open wells located 150 m away from the tank area. The open well and tube well provided water even in the summer month's which was not possible before the construction of water harvesting cum recharge tank.

It has been observed that immediately after the construction of the tank, the productivity of both the *kharif* and *rabi* crop increased. Similarly, the area in his field under wheat has increased tremendously than chickpea which is a clear cut indication that he is regularly getting sufficient irrigation water to grow wheat crop which require 2-3 more irrigation than chickpea. The yield enhancement was also influenced by the change in traditional crop varieties of wheat and chickpea due to availability of additional water for irrigation. Even in the year 2013-14, despite good monsoon rainfall (1339 mm), a continuous dry spell of 25 days in the month of September brought especially long durational varieties of soybean under severe moisture stress. The farmer provided a supplemental irrigation of 3 cm in two ha area to

Construction of Water Recharging Tank at Gaddukhedi (2010)
Technical Guidance to Line Department.

soybean variety RVS 2001-04 at this pod forming stage and this increased the soybean yield by 1.5 times than the other fields with same variety.

Before the construction of the tank, as the farmer owned four buffaloes and one cow, he was trying to earn extra income through dairy business

Effect of Waste Weir (Assessing Prospect of Storing more Water using the Existing Water Weir (Gaddukhedi).

After the Development of Sunken Pond (2011).

Different Stages of Sunken Tank Excavation at Gaddukhedi 2011.

but could not succeed as groundwater was not sufficient enough to grow fodder crops and even for drinking purpose of live stocks especially during post monsoon season. But after the construction of the tank and thereby due to increased surface and sub-surface water, he bought fifteen buffaloes in March 2012 and earning through dairying.

Making Advantage of Existing Waste Weir to Increase Water Storage for Recharge at Gaddukhedi 2011.

Another Wells in the Nearby Fields

The Nearby Well in June 2012

A Nearby Well in June 2012

A Nearby Well and Tubewell in September 2012.

Thus, it can be concluded that dovetailing of various schemes and participation by the farmer through land, labour or cost easily helped to develop a water storage body of sufficient size which not only retained huge amount of runoff but also recharged the nearby open well and tube-well. Further, this has reduced the amount of runoff which was otherwise leaving the watershed area without aiding to surface and sub – surface storage. The increased water availability also brought new area under high remunerative *rabi* crops and increased their productivity on sustainable basis.

Chapter 7

Tackling Dual Problems of Water Logging and Irrigation Water Scarcity through Natural Resource Management in *Nimar* Region

While working in the farmers fields through Operational Research Project for Dryland Agriculture, RVSKVV campus, College of Agriculture, Indore, it has been observed that many fields or their few portions are lying unused or underused for the various reasons related to natural resource management aspects. Further, it has been assessed that many fields which are being cultivated since long with lower productivity, can be reclaimed/ managed properly to enhance their productivity by tackling the site specific problems. The high runoff potential, scarcity of irrigation water, undulating topography, poor and restricted drainage system, temporary/permanent water logging, high erodible nature of soils, uncontrolled runoff are the major reasons for the lower productivity of both *kharif* and *rabi* crops resulting in abandoning of the cultivable lands or restricting its further use for higher production. The problem of each field is different and thus needs a site specific technological solution which requires scientific technical guidance for proper natural resource management. Bringing these areas into cultivation would certainly yield increase in the productivity of area through reclamation process. With this idea in view, the present study has been initiated during 2014-2015 in the village Limbai (Dist. Barwani) of Nimar region under Operational Research Project on Dryland Agriculture, College of Agriculture, Indore. The few fields of the village were facing with the typical problem of water logging during monsoon and early post monsoon season resulting in lower productivity of *kharif* crops and then delaying the sowing of *rabi* crops in time and even compelling to keep few fields *rabi* fallow due to higher moisture conditions. Further, non availability of surface and groundwater during *rabi* season also aggravated the situation resulting in lower productivity. Therefore, a study was initiated in 2014, in a farmer's field mainly to tackle/solve the dual problems of water logging and water scarcity through participatory and convergence approach to gain a firsthand experience in solving such typical problems as per the requirement of mandate of Operational Research Project on Dryland Agriculture.

The experimental present site of Operational Research Project on Dryland Agriculture is situated in village Limbai block Rajpur (Dist. Barwani) of Nimar region. In this village – Limbai block Rajpur (Barwani), in a meeting alongwith the line department officials and farmers of the village with ORP scientists held in September, 18th 2014, it was informed by a farmer Smt. Nathibai that despite owning 2.5 ha cultivable land in the village, she is not able to produce traditional crops satisfactorily even during *kharif* season. Some portions of her land do not allow interculture operations in *kharif* crops and crops are damaged/suffered due to heavy weed infestation. These cultivated fields remain in marshy condition till December month after the termination of monsoon and planting of *rabi* crops is delayed. Though, she owns an open well which remains full during monsoon season, it does not provide sufficient water to provide irrigation to *rabi* crops when it requires most. Thus in both the season, she is not able to harvest *kharif* and *rabi* crops like others located in her adjoining field. Because of overflowing of groundwater from her open well, adjoining area of about 15 ha was also getting affected and was producing lesser crop yields in *kharif* and *rabi* seasons. Immediately, it was decided by ORP team to visit her farm on the same day along with all line departmental officials and farmers. The team decided to assess the problems and suggest the probable solutions to agricultural related problem being faced by the farmer for increasing the crop productivity.

Assessment of the Problem

During the visit, it was observed that due to undulating topography and continuous sloppy field conditions, the soil is severely eroded with development of washes in various fields. Further, it was assessed that not only the fields are severely eroded but also water logging and stagnation of water on the fields damaging the crop growth and thus affecting the crop productivity. It was observed that the soil of the fields is clayey in nature and underlain by fragmented basalt (*Murrum*) and basaltic stones. The soil depth is also a limiting factor in few fields as soil depth varies from shallow to medium range. Further, it was observed that due to continuous groundwater recharge from ground soil strata, the *Kachha* open well located in the middle of the farmer's field gets fully recharged during monsoon. Due to nearby hilly terrain continuous groundwater recharge takes place during every monsoon season July month onward and the excess underground groundwater fills the well and flows/oozes out from the well and moves into to the lower area and creates problem of water logging in the cultivated fields severely affecting the *kharif* crops. It was revealed that the excess water from the well not only stagnates the fields but also remains in the fields till December end. This does not allow the farmer to cultivate the field with *rabi* crops. After December month, the water table suddenly goes down and the poor availability of water from the well does not make possible even *Palewa* (pre sowing irrigation) for *rabi* crops. In January, the open well totally dried up and remains dry till July end.

Similarly, due to overflowing of groundwater from the open well of the farmer was also affecting the cultivated fields of about 15 ha of other adjoining farmers. The both *kharif* and *rabi* crops in this 15 ha area was also suffering due to water logging and severe weed infestation resulting in lower yield levels. Also the conditions of excess moisture hampering the timely inter culture operation and timely sowing of mainly *rabi* crops in these fields.

Strategy and Solution of the Problems

After the spot visit and further discussion, it was assessed that with scientific and appropriate interventions, not only the dual problem of water logging and water scarcity can be tackled but also fertility of a large portion of the field can be increased. Thus, it was decided that to tackle the problem of water logging and stagnation, a suitable drainage system be provided so that the excess water drains out safely and the depressions and low lying fields be reclaimed to avoid water stagnation. Similarly, to increase the irrigation water availability it was decided that an excavated water harvesting tank be constructed which will collect the runoff water and the excess drainage water from the drainage system to avoid the water logging. Thus this will collect the huge amount of runoff and drainage water to be used during the limiting moisture conditions. Further, the excavated soil would be spread on the upstream side fields to make them level so that water stagnation in the field does not take place and water moves out the fields towards tank site without causing damage to crops. At the same time due to location of the tank, it would be possible to safely dispose off the excess water from the tank outlet and therefore avoiding the chances of water logging.

Water Recharging-cum-Harvesting Tank

A suitable site in the lowest portion of the field was selected considering all the hydrological aspects in September 2014 during the problem assessment visit along with line department officials and farmers and ORP team members. In this portion, technical guidance was provided to line department personnel and an excavated tank was developed using watershed funds of Rs.1.0 lakh available with line department in May 2015 using JCB machines and tractor trolleys for the transportation of excavated soil in the depressed field portions on the upstream side. The work was completed by providing suitable boulder inlet and outlet totally under the guidance of ORP team in May 2015. This tank of 3000 cu.m. capacity was allowed to fill with the runoff water in the following monsoon season. By using the excavated soil, a sort of diversion bund was provided on the upper portion of the adjoining fields of other farmers so that the runoff water from the upper hilly portion does not enter into the other farmers field and then into the farmers field with higher velocity to avoid the further chances of soil erosion and development of washes.

A depressed portion of the farmer's field on the upper side of well was elated using the excavated soil so that a sort of bedding system is developed to grow the crops like maize, cotton which are more susceptible to water logging.

Construction of Water Drainage System

It was observed that the water logging was taking place due to continuously recharged groundwater in the open well (during monsoon and post monsoon season till November month) which was coming out of the well and moving toward fields located on lower side of well laterally. Thus, it was planned that the excess water be drained out in the excavated water harvesting tank by connecting the well and tank by a subsurface drain. For this purpose, a PVC underground pipe line (7.5 cm dia.) of 150m length below 45 cm soil surface was laid from well to tank site so that the excess water coming into the well drains out safely in the tank area without overflowing the well and causing any water logging in the lower side fields due to water coming out of the well. Thus provision of draining the excess groundwater overflowing the well was made so that it is collected in the excavated water harvesting tank. The underground PVC pipe line from open well to water harvesting tank was so laid that it would not interfere with the normal cultivation operations in the fields. A lid was also placed at both the ends of pipeline to avoid entry of rodents *etc.* into the pipeline during dry period. The cost of drainage system (20000/-), which was installed under the technical guidance of ORP team, was borne by the farmer. Thus, for carrying out the work of exaction of water harvesting tank and installing sub surface drainage system to tackle the dual problem of water logging and water scarcity, the convergence and dovetailing of the funds from line department, technical support and guidance from the ORP side and farmer participation was ensured.

Effect of Water Harvesting Tank

The timely onset of monsoon season during June 2015 in the area, allowed the sowing of various crops in the farmers field. Further, two runoff events took place which filled the water harvesting tank in the month of June itself 2015. However, thereafter, a dry spell of about 20 days took place which brought the crops to moisture stress condition. Since, the open well was totally dry during June –July month, no groundwater was available for irrigation. However, the harvested water in the tank was utilized by the farmer for providing life saving irrigation mainly to *kharif* crops like maize, green gram and Pigeonpea. The monsoon revived again thereafter, and subsequent runoff events filled the tank to its capacity again.

A fifteen days dry spell was again experienced in September 2015 and this time also the harvested water was utilized for providing supplemental irrigation to *kharif* crops. However this time onward, the open well recharged to its full capacity and the provision of sub surface pipe system drained safely this excess groundwater into the tank to refill it again. After the harvesting of *kharif* crops, the harvested water was again utilized for providing *Palewa* in 0.5 ha area for the planting of *rabi* crops (Table 7.1).

Table 6.1: The Land Use Area of the Farmer Before and After the Construction of Water Harvesting Tank and Installation of Sub-surface Drainage System

Particulars	*Crops and their Conditions*				
Before the treatment	*Kharif Crops*	*Maize*	*Groundnut*	*Pigeonpea*	*Green gram*
	Area (ha)	1	0.5	0.5	0.5
	Weed	severe	severe	moderate	severe
	The yield levels were almost 1/3 of the normal yields.				
	Rabi Crops	*Wheat*	*Gram*	-	-
	Area	0.5	0.5		
	The yield levels were almost 1/3 of the normal yields.				
	About 1.5 ha was kept fallow due to water stagnation and limiting water in *rabi.*				
	About 15 ha area of other farmers was also getting affected due to water stagnation problem. Fields were heavily infested with weeds.				
	Irrigation from open well – 1 or 2 during *kharif* and only 1in *rabi*				
After construction of water harvesting tank and laying of subsurface drains (0.1 ha was used for tank construction)	*Kharif Crops*	*Maize*	*Groundnut*	*Pigeonpea*	*Green gram*
	Area (ha)	1	0.4	0.5	0.5
	Weed	Almost free from weed infestation.			
	Almost triple yield enhancement was recorded in *kharif* season in 2015.				
	Rabi crops	Wheat	Gram	Onion	-
	Area	1.5	0.4	0.5	
	No area was kept fallow due to water stagnation and limiting water.				
	Timely sowing of *rabi* crops was possible and condition of the crop was very satisfactory like other normal fields in November 2015.				
	No area of other farmers affected due to water stagnation problem and they are growing normal crops like cotton, green gram, onion, chilli, Pigeonpea, ground nut, maize, wheat and gram *etc.*				
	Irrigation from open well – 1 or 2 during *kharif* and only 1 in *rab*i. Irrigation from tank – 2 in *kharif* and 1 in *rabi*				

Problematic site

Open well in the field

Overflowing Open well in the field

Open well in the field – water oozes out

Fields remain wet throughout monsoon – damage to crop – more severe weed infestation

Fields remain wet throughout monsoon – damage to crop – more severe weed infestation

Water logged area – damage to kharif crops, delays rabi planting

Water logged area – delays rabi planting

Site Selection for water harvesting tank — LIMBAI 2014 — 18.9.2014

Site Selection for water harvesting tank — LIMBAI 2014 — 18.9.2014

Excavation for the construction of water harvesting tank — LIMBAI 2015 — 18.5.2015

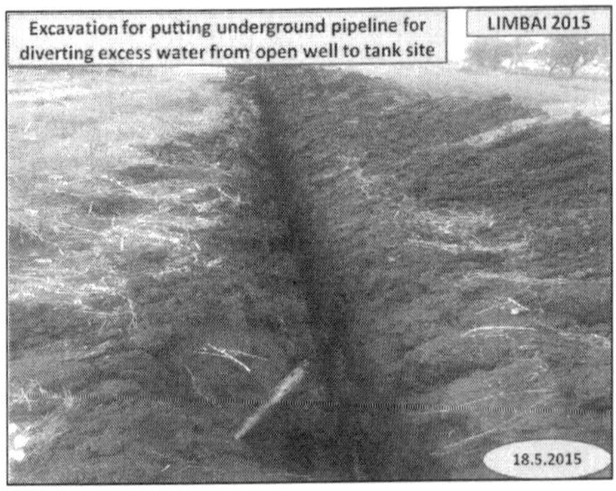

Excavation for putting underground pipeline for diverting excess water from open well to tank site — LIMBAI 2015 — 18.5.2015

Excavation for putting underground pipeline for diverting excess water from open well to tank site — LIMBAI 2015 — 18.5.2015

Pipe connecting open well and WHT — LIMBAI 2015 — 6.6.2015

Filling of water harvesting tank — LIMBAI 2015 — 26.6.2015

Fields ready for rabi planting – free from water stagnation — LIMBAI 2015 — 18.10.2015

Impact of Sub-surface Drainage System

As predicted, after the arrival of monsoon and continuous groundwater recharge thereof, the existing open well got fully recharged in the first week of August 2015. At this, time the lids placed at the both end of the underground pipe line was opened which safely drained out the excess groundwater otherwise overflowing the open well into the water harvesting tank. The immediate impact of this system was realized and observed that no water logging and water stagnation was observed in the fields located on the lower side of the well. This not only restricted the weed infestation in these fields but also saved the *kharif* crops from getting damaged due to marshy condition. This also allowed the timely inter culture operations and ensured better crop growth. Therefore, due to provision of sub surface drainage system the satisfactory *kharif* crops were harvested for the first time by the farmer which were otherwise almost one third of the normal yield levels. The sub surface drainage system allowed timely sowing of *rabi* crops in the otherwise fields with un-limiting moisture and condition of the crops was very satisfactory like other normal fields in November 2015.

The sub surface drainage system not only drained off the excess groundwater safely to protect the *kharif* crop getting damaged due to overflowing of open well, but also stored it in the constructed water harvesting tank. The additional stored water other than the surface runoff was thus collected in the tank and utilized for providing irrigation to both *kharif* and *rabi* crops which was otherwise never possible.

After the treatment, it was also observed that the 15 ha cultivated fields of other adjoining farmers also improved tremendously and thus allowing growing both *kharif* and *rabi* crops as per their satisfaction. No longer, these fields are suffering due to water logging and severe weed infestation resulting in higher yield levels of *kharif* crops. This has also allowed these farmers to carry out inter culture operation timely and timely sowing of *rabi* crops in these fields (Table 7.1).

It is concluded from the above study that assessing the site specific problem and then providing its solution scientifically can enhance the crop productivity in the fields otherwise lying unused or underused due to various reasons related to natural resource management aspects. In the present case too, a dual problem of water logging and water scarcity was tackled successfully by following principles of soil and water management through providing suitable drainage system and by constructing water storage structure in participatory mode by dovetailing and convergence process. Many site specific problems related to natural resource management still exist in the region and can be tackled scientifically.

Chapter 8

Success Story: Mitigating Adverse Climatic Conditions through Water Harvesting Tank in Farmer's Field

In *Malwa* region too, the early onset – late withdrawal, early onset – early withdrawal, late onset – early withdrawal, prolonged dry spells, continuous wet spells *etc.*, have been the features of the monsoon. Frost like situations in January- March, untimely rains, windy – hail storms also being experienced in the region since long. The climatic conditions have never been almost normal or favourable enough particularly for both *kharif* and *rabi* crops from the point of view of higher production. The limiting and non limiting moisture conditions caused by inadequate and uneven distribution of rainfall continue to be the most important climatic aberrations, which influence the agricultural production in this region. This has been the climatic trend of the region earlier also like most of areas of India. Therefore, producing the different crops with satisfactory yield levels without total damage to crops grown even under adverse climatic conditions should be the adaptation strategies for rainfed production system. Recently, excavation of farm ponds on the individual farmer's field has been found extremely useful in turning the mono-cropped area into multi-cropped/crop diversified area in this region.

Team of Operational Research Project for Dryland Agriculture, College of Agriculture, RVSKVV, Indore is involved with the farmers of the region and carrying out various scientific agricultural activities which may help the farmers to produce both *kharif* and *rabi* crops without any significant yield reduction even under very adverse climatic aberrations. Since soil and water management is highly critical for adaptation to climate change, the project made several attempts towards water conservation and water harvesting improvement in irrigation accessibility in the farmers' fields. To deal with the impact of climate change, team tried to enhance the water availability by constructing various water harvesting through farmers participation and dove tailing activities. Further, technical guidance were also provided to the farmers for the creation of water bodies in the farmers' fields itself for improving water and rain water management. A huge water harvesting tank was developed in a village Mordhan (Khudel) district Indore in May 2007 under the technical guidance of project team. The impact assessment study on various aspects of rainfed practices was made since then in the farmer's field after the increased water availability. The research experiences gathered by the team are, therefore, being reported for its wide scale adoption and developing site specific strategies.

It was observed that in this village, the scarcity of particularly surface water and groundwater in general, is the major constraint for growing satisfactory crop production. The combined family of the farmer owns about 27 hectare land and is growing mainly Soybean in *kharif* and Chickpea and wheat in *rabi* since 1995. He tried to dig up many tube-wells in his fields but was never satisfied with the quantity of irrigation water produced. It was clearly observed that the entire village was dependent on the rainfed farming practices. Even if the area gets good rainfall, the tube-wells provide water till December – January month only. Therefore, despite owing sizable cultivated field, the farmer's family was not able to produce higher crop production. However, it was observed by the team of Operational research Project for Dryland Agriculture, College of Agriculture, Indore that due to topographical features of the area associated with monsoon rains in the region and inherent soil physical properties (soil is clayey in nature with higher runoff producing properties), there is lot of potential for developing huge water harvesting tank in the farmers field itself. It was also assessed that since the field is located by the side of national Highway, there are chances to store additional runoff coming from other areas too in the proposed water harvesting tank. Thus, it was suggested that for irrigating 25 hectare atleast twice or thrice in *rabi* season, a huge water harvesting tank be constructed in his field. It was estimated that for this purpose, an area of one hectare is required and the farmer has to spend approximately Rs.10 lakh for the excavation and spreading of excavated materials into adjoining fields to make it leveled which are having several washes, undulating slopes due to continued runoff movement and soil erosion process within the fields. Considering all the prospects and aspects, the farmer readily agreed to provide one hectare land for this purpose and to bear the entire cost of developing the tank under the technical guidance of ORP team. Immediately, in March 2007, the excavation work was started using back hoe loaders machines and dumpers. The excavated soil was then spread into lower, undulating fields to make them in proper shape and size. The soil was also used to strengthen the existing entire farm road in the field. The work was completed in May 2007 with development of water harvesting tank in one hectare area with 80000 cubic meter storage capacity by investing approximately Rs.11 lakh. Since the natural outlet of the tank was located by the side of highway, the same was utilized as inlet for the tank so that after filling the tank the excess water flows out from the same drainage line. Thus, by keeping the common inlet and outlet of the tank, a sum of 50000/- were also saved. Immediately in the following monsoon of 2007, the tank was allowed to fill with the runoff water to the brim and huge amount of water was stored in it. Subsequently the stored water was utilized during dry spells in *kharif* and for pre-sowing and after sowing irrigation in *rabi* seasons. Since the constructed water harvesting tank is so located, it was observed that despite scanty or low rainfall year, the tank got filled to its full capacity since 2007 onward. Even after the use of stored water in *rabi*, the tank was refilled by the tube-well water to enhance the conjunctive use of groundwater and surface water.

Changes in Cropping Pattern and Crop Diversification

Prior to the construction of water harvesting tank, the farmer used to grow soybean in 25 ha in *kharif* and chickpea and/or wheat in *rabi* only in a limited area depending on the availability of water from existing tube-wells *etc.*, with low crop productivity. Prolonged dry spells during many rainy seasons, found to be main reason for the low productivity of soybean in *kharif*. Sometimes due to severe moisture stress condition even total failure of wheat was also reported. Earlier, the crops were also reported to be adversely affected by frost, windy storm and untimely rains during winter season. Even total failure of *rabi* crops have also been observed in many fields either due to severe stress or/and adverse climatic conditions and non availability of sufficient irrigation water.

After construction of water harvesting tank in 2007, it was observed that in 2007 – 08, the farmer grew wheat in 15 hectare and chickpea in 10 hectare in *rabi* by utilizing the stored water as pre-sowing irrigation and the following irrigations as per the requirement of the crops. It was observed that due to storage of runoff water in the constructed water harvesting tank, entire area of 25 hectare could be provided irrigation water which was otherwise not possible in any way with the existing tube wells. The farmer used both the submersible as well as diesel pumps for irrigation purpose from the tank. Therefore, despite the irregular electricity supply, by using both pumps could irrigate the entire 25 hectare land in sequence. This not only saved the time required for irrigating the field by 1/3 but also allowed the farmers to cultivate the entire field in time.

It was observed that generally in this region, the tube wells provide groundwater with reduced discharge after the termination of monsoon season as groundwater depletes at faster rate due to overexploitation. Thus, generally very few tube wells operate throughout the year otherwise most of the tube wells go dry or supply water only for 2-3 hours that too with very reduced quantity. At the same time electric power supply is a deterrent in providing continuous irrigation to field's crops. Sometimes the supply of tube well is so less that direct irrigation is not possible as water advancement process does not take place particularly in medium and deep clayey soils due to cracking pattern and irrigation process is not completed despite pump running for several hours. Therefore tube well water was collected in the tank for few days in this case and then the stored water was used for irrigating the nearby fields using a high discharge diesel/electric pump so that a larger area can be irrigated at a time. During 2007, it was also observed that the farmer used this tank as storage tank even after utilizing the entire stored runoff for irrigation. He refilled the tank by tube-well water without using it directly to irrigate the field crops. It was also observed that since the percolation rates of the tank is very meager, the farmer is collecting tube well water in the tank for its utilization for raising even summer/vegetable crops. Thus, the farmer relied more on stored water in the tanks than the tube well water as most of the tube wells dry up suddenly/abruptly due to groundwater depletion. Therefore, it was observed that the farmer not only got higher wheat production but also increased the productivity with this additional irrigation water. In the year, 2007-08, itself, the farmer produced 900 q wheat and 200 q chickpea which was otherwise almost the double than the normal condition. Therefore, it was estimated that the farmer generated an additional income of Rs 11 lakh with that additional and enhanced production of wheat and chickpea.

Introduction of New Crop Potato

The tank got filled with the runoff water in the month of June itself in 2008, as the runoff producing rainfall received in early monsoon. This year during *kharif*, the entire 25 ha area was brought under soybean cultivation with the combination of early and medium durational soybean varieties like JS 9560, JS 9305. Despite good initial monsoon condition, a dry spell in the month of August severely affected the soybean production in the adjoining fields. However, in the farmer's field, the harvested water in the tank provided a life saving irrigation (3 cm) to entire 25 ha soybean crop which enhanced its productivity by 1.5 times than the other fields. A runoff event in the month of September 2008 again filled the tank. Encouraged by the availability of huge surface water in the tank and last year's experience, in 2008-09, *rabi* season, after providing pre sowing irrigations, a new crop of potato was introduced after soybean harvest by the farmer on the suggestions of project team. Improved varieties of Potato *viz.,* Kufri lavkar, Chipsona, Kufri sindoori *etc.,* were planted in different fields/plots in sequence as per the availability of field and its proper condition for sowing. In this year, potato was sown in 7.5 hectare area. The other fields were sown with different varieties of wheat (GW-366, Lok-1, Harshita, Amrita, Pusa anmol, Pusa mangal, JW-17) and chickpea (JG-130, JG-16, JAKI-9218, JG-6, Vishal, JGK-3, KAK-2) crops in 8.5 ha each. After the harvest of potato in December month onward, the fallow fields were once again sown with wheat crop of different varieties made available to him through the project team till January 2009 last week after pre sowing irrigations in each harvested field in sequence. Thus all the fields were having wheat crop of different age, height and varieties. Thus during 2008-09, after the harvest of soybean, the farmer produced a new crop of potato along with wheat and chickpea. This year, all the field boundaries of the tank were sown with pigeonpea mainly for the domestic purpose.

De-silting and Deepening of Tank

In 2009, further deepening of tank was carried out and the excavated de-silted clayey soil was spread in the adjoining cultivated field to fill up the depressed portion and washes to increase the soil fertility status. For this purpose, again Rs.3 lakh were spent by the farmers for excavation and spreading of excavated soil in the field and its minor leveling. This ensured proper drainage from the cultivated fields and even chances of temporary water stagnation in it avoided. Since 2009 onwards and till 2015 the farmer is growing continuously three crops in *rabi* season with higher productivity and thus ensuring higher income level. Different varieties of wheat, chickpea and potato are being cultivated.

Minimized Losses during Adverse Climatic Conditions

In *Malwa* region, crop productivity suffers mainly on account of moisture stress towards the termination of monsoon and it needs supplemental irrigation. The stored water utilized for providing life saving irrigation to soybean crop during 2009, 2011, 2013 and 2014 during prolonged dry spells.

In 2013-14 due to water logging condition the soybean crop was damaged in the fields of many farmers in the area. However, the loss due to water logging was minimum in this field as most of the fields are having better drainage provision and proper leveling for the runoff collection in the water harvesting tank. Despite receiving higher rainfall than normal, a 25 days dry spell during September 2013 compelled the farmer to provide life saving irrigation (3 cm) to long durational varieties of soybean and this reflected in higher yield production. Thus both the adverse climatic conditions like water logging in the initial stage and moisture stress towards the termination of monsoon could be successfully tackled without any appreciable loss in crop yield. Similarly during *rabi* season the untimely rains, frost and hailstorm affected the crop productivity in other fields of different farmers. Since, in this 25 ha area after the potato harvesting the different fields was cultivated with wheat and chickpea at different time intervals by providing pre-sowing irrigations in sequence. Therefore at the time of untimely rains, frost and hailstorm the height, age group was totally different in various fields and this difference minimize the adversity of climate on crop productivity. Even at the time of harvest the untimely rains damaged the harvested wheat crop more than the standing crop in the field. A combination of wheat crop in different stages *viz.*, manually harvested wheat in bundles, wheat ready to harvest and wheat still in grain forming stage in the same field was possible only due to enhanced water availability which ensured adjusting the cropping sequence, including changing the timing of sowing, planting. Thus adverse impact of untimely rains and windy storm could be avoided and total failure to wheat was not observed in farmer's field like others.

Effect of Deep Ploughing

It was also observed that infestation of deep rooted weeds particularly Kans (*Sacchaurum spontaneum*), gajar grasss (*Parthenium* sp.) and mustard was very common in the region. The fields are, therefore, deep ploughed by the farmer using a tractor drawn reversible mould board plough initially provided by the project in 2007-2008. Encouraged by the result, the farmer himself purchased one tractor drawn reversible mould board plough and deep ploughing his fields once in a three year since then onward. This practice not only reduced the weed infestation but also increased the soil moisture profile in order to reduce the runoff and chances of water logging and further damage to crop. Similarly, it was observed that the expenditure towards the weedicides was reduced almost by one third. Breaking of hard layers below the root zone also ensured better and bold size of potato in deep ploughed fields than un-ploughed fields. This not only improved the soil environment but also ensured better germination of rainfed crops. Because of increased *in situ* conservation, this practice made possible, sowing of *rabi* crops with satisfactory germination without pre-sowing irrigation. The practice of deep ploughing also reduced the frequency of irrigations and saved precious irrigation water for its further use. It has also been experienced that deep tillage is another viable option to increase productivity mainly due to enhanced *in situ* moisture conservation in addition to providing the salutary condition for crop growth. Deep tillage was also found to contain weeds, which are instrumental in lowering the productivity of crops by way of offering severe competition to crop plants for natural resources.

Agronomical Practices followed to Address Climatic Adversities

Crop diversification was adopted by the farmer after the construction of tank which not only increased the cropping intensity but also enhanced the income of the farmer. Due to inclusion of potato in the cropping system, provided the farmer opportunity to cultivate wheat and gram in December – January month thus by avoiding impact of the high temperature on *rabi* crops growth during early stage and even save the crops due to low temperature during reproductive phase. The farmer also followed crop rotation system for saving the crops from getting affected due to fungal and bacterial infections on the advice of the scientists. Inter-cuture operation during dry spells particularly in *kharif* not only reduced the infestation of weeds but also conserved the soil moisture. Different potato planting systems *viz.*, one row and two rows planting on raised bed- furrow with introduction of drip irrigation system to enhance the irrigation water efficiency was also followed by the farmer. Due to enhanced water availability, the farmer is now irrigating the crops during occurrence of frost instead of traditional temperature increasing system of burning straw all around the fields.

Introduction of Micro-irrigation System

On the insistence of project team, the farmer started using drip irrigation system particularly for growing potato. This practice resulted in proper water management for irrigating crops with higher irrigation efficiency. This also reflected in higher crop production of various crops grown by the farmer. The rain gun and sprinkler irrigation systems were also utilized by the farmer for irrigating wheat crop particularly. Sometimes, he used sprinkler for irrigating soybean crop during dry spells. Thus, increased water availability also not only encouraged the farmers to adopt crop diversification practices but also motivated him to adopt micro irrigation for better irrigation water management for higher efficiency and water saving. But it is very clear that particularly in *Vertisols* of this region, better crop germination and growth can be ensured only by providing flood irrigation through border method before or after the sowing. For the subsequent irrigation, sprinkler irrigation can be used.

Even the use of *Livocin* a growth regulator restricted the unnecessary biological growth in wheat but also strengthen the roots. Due to this the least damage was observed during adverse climatic conditions like untimely rains, hailstorm and frost because of restricted plant height of wheat crop. This also reduced the chances of damage due to lodging. Therefore, overall damage in comparison to other farmers was far less and almost 1.5 times more yield was recorded in wheat crop due to use of this growth regulator.

Khudel 2015

Santosh tank
130*100*7.5 m
KHUDEL 2013
Santosh followed Soybean- Potato- wheat sequence
KHUDEL 2013
CROP DIVERSIFICATION DUE TO ENHANCED WATER AVAILABILITY

Matured crop
Harvested crop
Khudel 2014
25.3.2014
25.3.2014
Wheat crop in March 2014
25.3.2014
25.3.2014

Khudel 2014
Multiple wheat and gram varieties
21.1.2014
21.1.2014
DIFFERENT CROPS GROWN IN RABI SEASON 2013-2014
Crop management for mitigating weather aberration
21.1.2014
Different Age and height minimized the loss due to heavy rains, hail storm and wind speed in January 2014
21.1.2014
21.1.2014

Khudel 2014
Wheat planting
21.1.2014
21.1.2014
Cultivated third crop using harvested water
21.1.2014
21.1.2014

Khudel 2014
Use of reversible plough for deep ploughing a regular practice once in three year

Khudel 2015
Wheat crop in March 2015
18.3.2015

More weed infestation in unploughed field
NINGNOTI 2015
5.2.2015

Thus in the present study, the adjustment in the cropping sequence, including changing the timing of sowing, planting, spraying, and harvesting, which was possible due to enhanced water availability in the form of surface water helped to take advantage of the changing duration of growing seasons and associated heat and moisture levels. Altering the time at which fields are sowed or planted also helped the farmer to regulate the length of the growing season to better suit the adverse climatic conditions Farmer adaptation can also involve changing the timing of irrigation or use of other inputs such as fertilizers. A huge water storage structure created in the present study brought visible changes and helped in managing the natural resources in the field. The use of heavy machines not only completed work in time but also saved the expenditure, which would have been otherwise almost double if the manual force was utilized. Therefore, it can be pointed out that besides man, money, materials, machines also play a very significant role in natural resource conservation programme and thus its use particularly for developmental work should be encouraged.

It can be, thus, pointed out that due to increased water availability through the development of water harvesting tank, many adaptation strategies *viz.* introduction of tolerant cultivars resistant to temporary water logging and drought, modifying crop management practices, improving water and rain water management, adopting new farm techniques such as crop diversification particularly to deal with the impact of climate change could be adopted successfully to minimize the damage to crops due to adverse and changing weather conditions. Therefore, increasing irrigation water availability through water harvesting tank could be a certain way to mitigate adverse climatic conditions in *Malwa* region. Since Soil and water management is highly critical for adaptation to climate change, serious attempts towards water conservation, water harvesting improvement in irrigation accessibility and water-use efficiency essential for crop production should be made on priority basis. At the same time, the farmers have to be motivated for adopting on-farm water conservation techniques, micro-irrigation systems for better water-use efficiency, selection of appropriate crops, *etc.*

Index

I

K

L

M

N

O

P

R

S

T

U

W